50 THINGS YOU NEED TO KNOW ABOUT

PERIODS

Know your flow and live in sync with your cycle

CLAIRE BAKER

PAVILION

Contents

A note from your friendly neighbourhood period coach   4

Menstruality 101

#1   Hands up if you've ever complained about your period   9

#2   The difference between your period and your menstrual cycle   10

#3   What's a 'normal' period?   11

#4   Get in touch with your menstruality   12

#5   You and your period   14

#6   My cycle (kind of) looks like…   16

#7   Hormonal happenings   20

#8   The seasons of your cycle   22

#9   The BIG 'O'   24

#10   How to figure out where you are in the cycle   25

#11   How to chart (and sync) your cycle   26

#12   Winter observations – around cycle day 3 you might feel…   30

#13   Spring observations – around cycle day 8 you might feel…   30

#14   Summer observations – around cycle day 13 you might feel…   31

#15   Autumn observations – around cycle day 25 you might feel…   31

Winter

#16   Winter phase: menstruation   34

#17   Winter's superpowers   37

#18   Winter's vulnerabilities   40

#19   Blood and body love   42

#20   Tips for period problems   44

#21   Use a menstrual cup   46

#22   How to use your menstrual cup   47

#23   Reusable pads and period undies   48

#24   Winter cycle charting Qs   50

#25   Winter cycle-syncing tips   52

Spring

#26	Spring phase: pre-ovulation	56
#27	Spring's superpowers	58
#28	Spring's vulnerabilities	61
#29	How to make a less and more list	62
#30	Spring cycle charting Qs	63
#31	Spring cycle-syncing tips	64

Summer

#32	Summer phase: ovulation	68
#33	Summer's superpowers	70
#34	Summer's vulnerabilities	73
#35	Write an ovulation gratitude list	74
#36	Summer cycle charting Qs	76
#37	Summer cycle-syncing tips	78

Autumn

#38	Autumn phase: premenstruum	82
#39	Autumn's superpowers	85
#40	Autumn's vulnerabilities	88
#41	Write an Autumn list	91
#42	Self-care for PMS	92
#43	Autumn cycle charting Qs	94
#44	Autumn cycle-syncing tips	95

Period Positive

#45	Get to know your crossover days	99
#46	How to talk to your people about periods	102
#47	What does the menstrual cycle have to do with the moon?	105
#48	As to Qs I get asked a lot	106
#49	A menstrual manifesto	108
#50	Period positive resources	110

A note from your friendly neighbourhood period coach

Ever stuffed a tampon up your sleeve on your way to the office bathroom? Avoided eye contact with the cashier as you paid for your sanitary pads? Cursed your premenstrual moods while curled up with Netflix, a box of tissues, and a block of Cadbury's finest? Gritted your teeth in frustration at your body and how wildly inconsistent it seems to be? Felt overwhelmed, exhausted, and annoyed by your hormones? Wished you simply didn't have a period at all? Well, you are not alone.

Our society doesn't celebrate the menstrual cycle. It's all very hush-hush. We're told not to talk about it in public, brought up calling it 'the curse'. Even saying the word 'period' out loud is still a bit taboo, right? It's that 'time of the month' when 'Aunt Flo is visiting' and you've 'got the painters in'. My personal favourite… 'shark week'! But jokes aside, the truth is that it can be bloody hard living in a body that bleeds once a month.

So why a book about periods then?

Well, you see, they're the missing key. You may be surprised to learn that, if you're someone who menstruates, you are at this very moment in time (and at every minute of every day) experiencing one of the four phases of the menstrual cycle. You might also be surprised to hear that these four phases vary hormonally, which might explain why you can feel like a different person from week to week. The rise and fall of your hormones (like oestrogen, progesterone and testosterone) can affect mood, motivation, focus, energy levels, memory retention, cognition, desires, confidence, personality, libido, and how your body responds to stress. So, you know, quite a lot…

A bloody good thing

As your friendly neighbourhood period coach, I've collected together 50 bits of advice and know-how to help you understand this internal rhythm that you move through each month(ish). This will give you all the tools you need to work with your body, rather than pushing against it.

As you journey through this book, you'll learn…

- Why your period is positive and how to work it!
- That you're not crazy. It's 100% normal to have fluctuating energy levels, libido, cravings and, erm, personality traits, all month long.
- All about the four distinct hormonal phases (or 'seasons' as we'll call them) in your menstrual cycle.
- How to chart your cycle so you can identify your unique superpowers and vulnerabilities in each phase.
- How to find your flow by syncing your life, where possible, around each phase of your menstrual cycle. **Spoiler: what works one week, won't necessarily work the next.**

You'll also discover some fun facts (like what that wet stuff in your undies is and what your menstrual cycle has to do with the moon), and find some super savvy suggestions (like how to practise cycle self-care and how to chat to important people in your life about periods). You'll even pick up some excellent real-life insights from a few of my lovely clients — just a handful of the amazing women I got to know when leading workshops, running one-to-one sessions, and teaching courses (both online and IRL).

You'll soon learn that there is so much more to your period than pain and PMS, though of course we'll get to them too. By paying attention to your body and learning how to live and work with your cycle in an optimal way, you will be able to achieve the things you most want in your life with greater ease than ever before.

I'm aware that it's likely you've never been handed a guidebook to your menstrual cycle, or told that your period is a positive thing to treasure. This book is not a clinical or overwhelming guide to the menstrual cycle. What you'll find in the following pages is information that is easy to understand and a joy to implement, or rather to *experiment* with. That's the key word here. Consider this book an invitation to experiment with a whole new way of living and bleeding.

It's my hope that you'll use the information here to crown yourself as the authority in your own life. Because guess what world? Women bleed. You can tax our tampons, demonize PMS, use weird blue liquid in our pad adverts, and tell us to suck it up, but we're not about that life anymore.

We're period literate, baby — and that really is a bloody good thing.

Menstruality
101

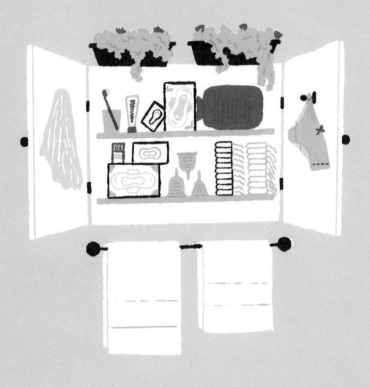

#1

Hands up if you've ever complained about your period

I always get a room full of raised hands when I ask this before teaching a workshop. Maybe you experience period pain. Maybe you find the whole thing inconvenient and unnecessary. Maybe you've never really considered the idea that your period could be anything other than a burden. Maybe you're trying to conceive and your period sadly announces, 'not this month'. Whatever brings you to this book, know this:

You are definitely not alone.

If you have complex feelings towards your period, I hear you. And yet, here you are — inviting in the possibility of period power and a positive relationship with your monthly flow. For that, I salute you.

#2
___

The difference between your period and your menstrual cycle

It's a common misconception that a period and a menstrual cycle are the same thing. Or that there's menstruation, the week of bleeding, and the rest of the time is same-same. Actually, if you're somebody who menstruates, then you will always (at every minute of every single day) be experiencing one of the many phases of the menstrual cycle. Your period is just one of those phases.

Your menstrual cycle is the entire inner process designed to create a pregnancy during your body's reproductive years. It's counted from the first day of your period to the day before your next period begins.

There are two key events in the menstrual cycle: menstruation (the release of blood) and ovulation (the release of an egg). These two poles are governed by daily hormonal changes that occur in your reproductive system. Say a big hello to your ovaries, fallopian tubes, uterus, cervix, vagina, vulva, mammary glands and breasts! They're all part of this process.

The menstrual cycle can be divided into four hormonal phases or 'seasons', which we'll get to shortly (see #8). It can also be looked at as a cycle of two halves. The first half of your cycle is called the follicular phase. It takes you from menstruation (the beginning of the cycle) right up to ovulation. The second half of your cycle is called the luteal phase. It takes you from ovulation back down to menstruation, where one cycle ends and a new one begins.

#3

What's a 'normal' period?

Periods can be affected by a myriad of things: changes in diet, sleep patterns, sexual activity, travelling (long haul flying always seems to mess with menstrual cycles!), getting an exciting but stressful promotion, how often you've been going to the gym, the list goes on...

Paying attention to your period over time will help you to understand what's normal for you. However, it is good to know the ranges that a typical, healthy period falls in. If you're an adult with a natural menstrual cycle (and you're not using any form of hormonal contraceptive or hormonal IUD), then a normal period plays out like this...

- Arrives every 24 to 35 days.
- Bleeding lasts between 2 and 7 days.
- Menstrual blood loss is between 5ml and 80ml (1 to 13½ tsp) over the entire period. A regular tampon or pad holds about 5ml (1 tsp) and a super tampon or pad holds about double that. Menstrual cups hold between 15ml and 30ml (3 to 6 tsp), depending on the size.
- Blood is a rich red colour and may change day by day, possibly being darker at the beginning and end of your flow.
- Small amounts of clotting or clumping for the first few days, or none at all.
- Clots are no bigger than 2.5cm (1in).
- No severe pain. Some sensation or discomfort is okay.
- Minimal menstrual issues such as headaches, breakouts, sore back, and sore breasts.

#4

Get in touch with your menstruality

If you get periods, it's likely you'll have about 450 in your lifetime, over about 35 years. Your menstruating years begin at menarche, which is the name for your first period, and end when you enter menopause. These are your cyclic years. Consider the timeline below. Where are you currently situated in your cyclic years? Are you closer to the beginning, approaching the end or somewhere in the middle?

Wherever you are, this female life journey, from menarche through to menopause and beyond, is referred to as 'menstruality'. Menstruality! Such a great word. Coined by New Zealand psychotherapist Jane Catherine Severn, you can use this word to describe your embodied experience of having a menstrual cycle.

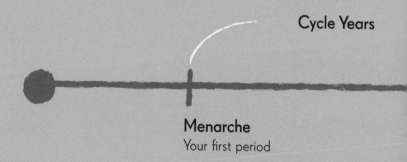

Cycle Years

Menarche
Your first period

Living in a world that is very keen on linear models of productivity and output can be tough for those of us with menstrual cycles. In fact, the 9-to-5 working day is much more suited to the male body's daily hormonal rhythm. Menstruality gives us a word to describe the experience of being in a body that ebbs and flows every month. Getting in touch with your menstruality is all about…

- Paying attention to your period.
- Observing how your menstrual cycle affects how you feel each day.
- Identifying your unique strengths and weaknesses (your superpowers and vulnerabilities) in each phase of your menstrual cycle (see #11).
- Being mindful of where you are currently situated in your cyclic years.
- Where possible, living in flow with your natural rhythm.

Menopause
Your periods stop

Perimenopause
Cycles start to change

#5

You and your period

Journaling is a simple but powerful method of uncovering the feelings, beliefs and ideas you have about your period, and also about yourself. What comes up when you ponder these writing prompts?

My first period was... _____

_____

Growing up, I felt my body was... _____

_____

My family taught me that my period is... _____

_____

Words I use to describe having my period are... _____

_____

Words I use to describe my 'vagina', 'vulva', 'pussy' are... _____

_____

The way I talk about my period with others is... _____

_____

My blood is... _____

_____

My loved ones' attitudes to my period are... _____

_____

When I get my period, I feel... _____

_____

Society tells me that my period is... _____

_____

Words I'd like to use to describe the experience of having a menstrual cycle are... _____

_____

To begin to appreciate my period I need to let go of... _____

_____

A kind approach to having a menstrual cycle might include... _____

_____

Very well done! What have you learnt about yourself?

#6

—

My cycle (kind of) looks like...

Your menstrual cycle experience might look and feel a little different to mine, but I wonder, does any of this sound familiar to you?

DAY 1
That first sign of blood feels like sweet relief after the intensity of the past week! Unless it's accompanied by cramps, in which case, not as fun. If I'm prepared, I'm ready with books by the sofa, a heat pack and snacks. If my period catches me by surprise, I just rest when I can.

DAY 2
Happy to chill and let my body do its thing. Not as happy to cook or clean or attend to responsibilities of any kind. Of course, life happens. And so I move slowly. When I relax, I enjoy the feeling of releasing and resetting. I mostly feel blissful and calm... but can also feel tired and bleurgh!

DAY 3
Hello little spark in energy! Also: sex dreams. My breasts and belly are reducing in size. I feel powerful – the 'period ideas' (you'll learn more about these in #17) have begun!

DAYS 4, 5, 6
Time to plan for the cycle ahead! I'm more energized, clear on my intentions and ready to get things moving.

DAY 7

Farewell, period cave! I feel light and free in my body and my skin is simply excellent. I'll wash and blow-dry my hair, book in for a wax, paint my nails and bounce my way through the day. High chance of mini skirts.

DAYS 8, 9

Feeling sassy and motivated, with energy and enthusiasm for work. Time to put those 'period ideas' into action. Also: has there been an increase of attractive people in my neighbourhood in the past week? Everyone is looking at least 20% more beautiful...

DAYS 10, 11

I am a real-life superwoman! I'm feeling on top of my life and I'm getting things done.

DAYS 12, 13, 14

Oh yeah baby, this is where it's at for me. I'll start to feel my body ovulating: a twinge in my lower belly, 'egg white' fluid in my undies (see #9) and a desire to get it on.

DAYS 15, 16

I'm a communication queen. Full of empathy and love, I tell everyone I know how much I appreciate them! I'll be cooking dinner for pals, hanging on social media, teaching workshops, calling my family, seeing clients and enjoying my (finite) extroversion.

DAYS 17, 18, 19

Ah hello there, little dip in energy. Things are shifting. Need a tad more sleep. Need a tad more time alone.

DAY 20

Feeling more sensitive, but still super productive – as long as I have lots of space and get to do things on my own timeline.

DAYS 21, 22

If I'm aware of where I am in my cycle and can be kind to myself, I like the sensation of slowing down, tuning in and feeling reflective. I keep exercising as it helps me to move through this change in energy. Prone to self-criticism and self-doubt.

DAYS 23, 24

Craving sweets and sugar! Can't stand rude people. Crying at TV adverts for life insurance. Very mad at any last-minute changes in plans. Baths are a very good idea.

DAYS 25, 26

I'm fabulous at editing and adding depth and emotion to projects. High chance of 90s Alanis on my Spotify. Want to eat all of the food. Inner critic still very much in the house. I'll prep for my period by making 'me time' in my schedule, decluttering the house, asking for help, prioritizing tasks — and stocking the fridge!

DAYS 27, 28

Need to dance. Need to sleep. Need hugs. Need to be alone. Need to urgently edit every bit of copy on my website. Need chocolate. Need to bleed!!!

DAY 1

Ah, sweet release! And also: hello sofa, my old friend. And the cycle begins all over again...

#7

Hormonal happenings

Hormones are the chemical messengers that regulate the behaviour of your organs. They're responsible for the quality of sleep you had last night, your mood, digestion, libido, whether you'll have energy for the gym tonight, and what foods you'll be craving at 4pm. Here are the key hormonal players that regulate your menstrual cycle...

Oestrogen

Kicking things off in the first half of your cycle, oestrogen can make you feel confident and turned on. This hormone prepares your body for ovulation and thickens the lining of your uterus. Oestrogen levels peak when an egg is released at ovulation and drop off in the second half of the cycle. As well as plumping up your skin and hair and increasing sexual desire, healthy levels of this feel-good hormone do wonders for our mental and emotional health.

Progesterone

Taking the reins in the second half of your cycle, progesterone production begins once ovulation has occurred. Progesterone is all about supporting a potential pregnancy by nourishing the lining of your uterus. The most *hygge* of your sex hormones (think cosy nights in with candles and hot chocolate), progesterone is soothing and can aid sleep. Consider it your 'adulting' hormone, nudging you to take better care of yourself. Once progesterone production halts in the luteal phase, you'll begin to bleed.

Hormone changes in an average cycle

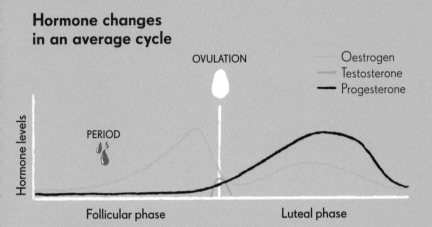

Follicular phase Luteal phase

Testosterone
Peaking in tandem with oestrogen at ovulation, this hormone is all fire. Increased strength? Tick. Better energy levels? Tick. A surge in drive, determination and a desire to win? Tick, tick, tick. Testosterone is important for bone strength, the development of muscle mass and contributes to an overall sense of wellbeing. Excitingly, it heightens sexual response and orgasm, too.

Luteinizing Hormone (aka LH)
A firecracker of a hormone, LH is present for only a tiny moment in the cycle, but it's got a very critical job to do: getting ovulation up, up and away!

Follicle Stimulating Hormone (aka FSH)
After your period ends, your ovaries do their pre-game stretches and warm up. The follicles in your ovaries grow nice and plump, and eventually one ovary will release an egg in your ovulatory phase. It's FSH that encourages and stimulates the follicles to develop into their full glory – so one ovary is all set to ovulate when the time is right.

#8

The seasons of your cycle

The stages of your menstrual cycle mirror the seasons of the year and, because they are so familiar to most of us, I like to use the four seasons to explain the four hormonal phases of the cycle.

The Winter of your cycle is your period. This is menstruation, when your body releases the last cycle's uterine lining. Levels of hormones are low. Just like on a cold, rainy winter day, you might feel like chilling indoors under a blanket in this phase, and that is A-OK.

After menstruation ends, you'll move into the Spring phase of your cycle. This is the pre-ovulatory phase. As oestrogen rises, you're blossoming and building up to ovulation here. Say farewell to the period cave because hormones are kicking off and you are back in the game!

A few days before you ovulate, you'll move into the Summer phase of your cycle. Ovulation is the release of an egg from one of your ovaries and it's a key moment in the menstrual cycle process. Both oestrogen and testosterone levels peak and you'll stay in this sassy Summer phase as you ovulate, and for a few days or so past ovulation.

You'll then move into the Autumn of your cycle – the premenstrual phase. Progesterone steals the show here, but not before having a 'hormonal dance off' with oestrogen a few days after you ovulate. Compared to the linear increase of hormones in Spring, Autumn is more dynamic. You're moving back towards menstruation. After progesterone levels decline and halt entirely, you'll cross back into Winter and begin to bleed again.

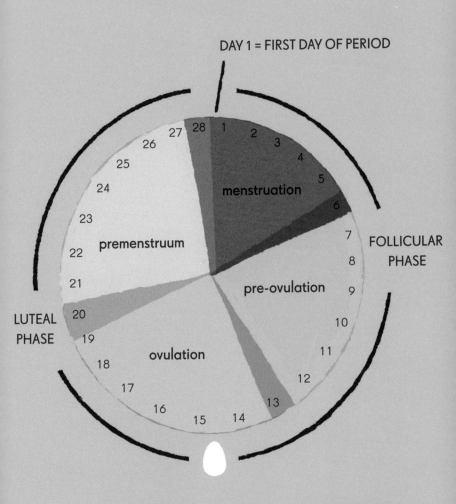

DAY 1 = FIRST DAY OF PERIOD

menstruation

premenstruum

pre-ovulation

ovulation

FOLLICULAR PHASE

LUTEAL PHASE

1 2 3 4 5 6 7 8 9 10 11 12 13 14 15 16 17 18 19 20 21 22 23 24 25 26 27 28

This chart shows what a 28-day cycle could look like, but every cycle is different. If your cycle is longer or shorter than 28 days, remember that you'll experience the timing of your seasons slightly differently.

● WINTER
○ SPRING
SUMMER
● AUTUMN

#9

The BIG 'O'

You guessed it: ovulation. The release of an egg from one of your ovaries. You can observe ovulation in your Summer phase by tracking both your temperature and your cervical fluid... the wet stuff in your knickers!

Charting Basal Body Temperature (BBT)

Progesterone (the hormone that kicks in post-ovulation) raises your temperature. Over a few cycles you'll begin to notice that your temperatures are lower in the first half of your cycle and slightly higher in the second half. To chart BBT you need to take your temperature each morning, orally, before you get out of bed, with a digital basal body thermometer. If you chart an increase of 0.2-0.6°C for three days or more, it's likely you've ovulated. You need to chart three lower temperatures followed by three higher temperatures (reflecting this increase) to confirm ovulation.

Charting cervical fluid

As your hormones shift day to day, so will this fluid. You know how some days you feel completely dry down there but on others you feel really wet? These observations hold a key to understanding exactly where you are in the cycle. If you chart this, it might look and feel something like this...

Winter = As bleeding slows, you'll notice some moistness or you'll feel dry.
Spring = Starts off dry or moist, becoming more wet and watery.
Summer = A strong feeling of wetness. Fluid might look clear, slippery and stretchy, like raw egg white. This is fertile cervical fluid and it's a clear sign ovulation is on the cards. Cervical fluid will dry up quickly after ovulation.
Autumn = Dry feeling continues. Any fluid looks white, cloudy and creamy.

#10

How to figure out where you are in the cycle

Have a look back in your calendar and find your last period. Now, the first day of your period doesn't include spotting. That first day of juicy red blood flow? That's the first day of your period, and the first day of your menstrual cycle. That's your cycle day one. Count forward from there and ta-dah! You've figured out today's cycle day. Remember, the next time you have a period, you'll start counting again from day one.

For someone who has a 28-day cycle, each season will be about 7 days, and ovulation will happen somewhere around day 14. This will vary because there are many factors that can affect when ovulation actually happens in a cycle. If your cycle is shorter than 28 days, it's likely you're ovulating earlier. If your cycle is longer, you'll be ovulating later in the cycle. This is because the luteal phase (the days between ovulation and your next period) is actually quite predictable. You can generally expect your period to arrive 11 to 17 days after you ovulate, but the days leading up to ovulation (the follicular phase) are more varied.

Simple season summary
- You're in your Winter phase when you have your period.
- You're in your Spring phase if your period has ended but you haven't yet ovulated.
- You're in your Summer phase if ovulation is about to happen, is happening, or has just happened.
- You're in your Autumn phase if you're post-ovulation but haven't started your period.

#11

How to chart (and sync) your cycle

Cycle syncing is about approaching the important aspects of your life differently in each phase of your cycle – because how you feel changes from season to season. It's less about micro-managing our schedule, and more about holding ourselves in the awareness that the hormonal journey means we shift and change from week-to-week. By gaining insight into our cycle, the way we approach everything from our attitude, creativity and productivity to our relationships, self-talk and self-care, can ebb and flow with these inner changes.

You need to commit to consistent daily charting to find the most golden cycle insights, but checking in doesn't need to take more than 30 seconds a day. But how do you chart your cycle? First you need to check in each day with your 'four bodies'...

Physical
How is your physical body today? Think energy levels, sensations in the body, cravings, digestion, appetite, libido, strength and endurance, and sleep patterns. What's your body telling you it needs today?

Mental
How is your head today? Think quantity and quality of thoughts, cognitive function, memory, motivation, insight, focus, attitude and perception. What do you notice about your mindset today?

Emotional

How is your heart today? Think mood, feelings, self-love levels, stress levels, confidence and self-esteem. Do you have a stable emotional landscape or do things feel more up and down?

Spiritual

How is your spirit today? Think gratitude, intuition, creativity, open-heartedness and levels of fulfilment. How connected do you feel to yourself or to something greater than you?

Once you've checked in with your four bodies, you need to think about your superpowers, vulnerabilities and self-care...

Superpowers and vulnerabilities

It's normal to have a place in the menstrual cycle where you feel 'at home' and more like yourself. It's also likely you'll have spots where you feel more vulnerable or uncomfortable. As you chart your cycle, pay attention to your unique 'superpowers' in each season: what are your strong points and positive qualities in each phase? Same goes with your vulnerabilities: what are your particular sensitivities or weak points in each phase?

Self-care practices

Finally, think about the self-care practices that will help you to optimize your strengths in each phase and take care of your vulnerabilities. Remember to keep asking yourself: how can I practise self-care today?

Then try out these methods to chart your cycle:

- **Use words.** Keep it simple. Write just one or two words for each 'body' that captures how you feel today.

- **Use an app.** If the convenience of digital tracking appeals to you, there are plenty of free period apps available. Choose one that allows you to input a variety of physical, mental, spiritual and emotional observations.

- **Use journaling.** If you already use a daily journal, add your cycle day next to the date and simply add your thoughts and feelings as you would normally. After a few cycles of doing this, flick back to notice patterns.

- **Use a notebook.** Divide each page into four sections. The first page of the notebook will be for day 1 observations, the second page will be for day 2 observations, the third page will be for day 3 observations... and so on. Use one section per cycle and you'll end up with four cycles worth of observations — this will make it easy to notice patterns.

- **Use the notes app on your phone.** Create a new note for each day of your cycle and jot your observations down. With each new cycle, go back and start again using the same note page for each day in the cycle. Another easy way to notice patterns over time!

- **Use my menstrual map.** Using one map per cycle, jot down your daily observations for each season. (To download your own map, see #50.)

Okay, you're nearly ready to explore the four seasons of your menstrual cycle. Are you excited? I hope so. Before we dive in, overleaf is a selection of common seasonal observations.

Your Menstrual Map

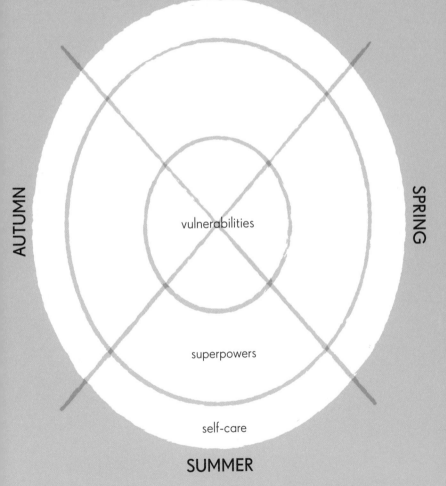

WINTER

AUTUMN

SPRING

vulnerabilities

superpowers

self-care

SUMMER

As you begin to chart your cycle, you might make observations like…

#12

Winter observations – around cycle day 3 you might feel...

- a slight surge in vitality, like there's renewal running through your veins.
- or maybe still a little sleepy, dreamy, floaty.
- or possibly (especially if you've been busy) like you want to dig a hole into the earth and just lie there a wee while.
- relieved, connected, open-hearted, powerful, exhausted, rested, sensual, sad, strong. There are no rules when it comes to menstruation.

#13

Spring observations – around cycle day 8 you might feel...

- surprised when you wake before your alarm. Sleep? Suddenly it's not as tempting as it has been for the past two weeks.
- a tingle between your legs, like something sweet is stirring.
- inspired to borrow books from your library. There's so much to learn!
- like washing your hair, waxing your legs, fluttering those lashes and flashing some skin. Oh, and a little online shopping treat sounds lovely.

#14

Summer observations – around cycle day 13 you might feel...

- as if everybody you pass in the street has suddenly become much, much more attractive.
- like today is the day to take that stand-up-paddle-boarding power pilates class you've been eyeing up. Who doesn't love crushing their core while floating on water?! Get it, girl.
- that in fact, you do have your life sorted out! You're nailing it!
- just quite up for a shag, really.

#15

Autumn observations – around cycle day 25 you might feel...

- like eating four burgers, an entire watermelon, three slices of birthday cake and a curry.
- exceptionally irritated with people talking in the quiet carriage on the train.
- like your dreams are so weird they're straight from a Wes Anderson film.
- like crying at life insurance ads.
- really, really, really ready to bleed.

Winter

#16

Winter phase: menstruation

Let's start with your period, or menstruation, or your moon time, or lady week, or whatever you like to call this phase. Your Winter is the beginning and end of your menstrual cycle. I wonder, does it feel more like an ending or a beginning to you?

We all experience menstruation differently, and over the course of your reproductive years it's normal to notice some changes in your period and how you feel about it. If the premenstrual week (your Autumn) is tricky for you, it can feel like sweet relief when your period (your Winter) arrives. But for many people, this phase typically isn't a time of celebration. If you experience pain, cramping or discomfort when you bleed, then your period might feel like an annoying inconvenience or even debilitating.

But what if I told you that your period isn't actually a curse, but a source of power? There are many reasons why menstruation is a gift. When you understand how to work with your period superpowers, this week can be a time to reset and replenish your body with gorgeous self-care practices, and to get clear on your priorities. Learning how to slow down and nourish yourself when you're bleeding has many benefits. You may even begin to look forward to your period – seriously!

THE WINTER LOWDOWN

When is your Winter phase?
Approximately cycle days 1 to 6 (or it may start a day before bleeding).

What's your body doing?
Both oestrogen and progesterone are at rock bottom, the lowest levels you'll experience for the entirety of your cycle. Combine that with the fact your uterus is involved in the process of releasing its contents via your vagina and it makes sense that energy levels may be low.

Towards the end of your Winter phase oestrogen begins to rise steadily, which is why you might feel a little spark in energy and clarity around cycle day 5 or 6. The new cycle has begun.

#17

Winter's superpowers

Release and reset

Who doesn't love the feeling of a fresh start? Healing after heartbreak, the first day of a new job, moving to a new city where no one knows your name. It's the sweet feeling of starting over, and your period can feel just like this! Your hormones have wrapped up one cycle and are about to begin a brand new one, so the Winter phase provides a fantastic opportunity to begin again and rebalance your system.

As your body physically releases what it has built in your womb over the past month, take a moment to think about what you'd like to release along with it. Crappy self-talk? Your 4pm sugar habit? That babe on Tinder who hasn't replied for days? Let it all go with your blood. You might feel the effect of this 'release and reset' in the size and shape of your belly and breasts, in your energy levels and libido, or in your emotions, thoughts and attitude. Work with this superpower by being clear on what you're releasing and identifying what this new cycle is going to be about for you.

Recharge

Rest and recharge at menstruation? But isn't 'period positivity' and women's empowerment about our periods and female physiology *not* holding us back? We can all agree that no one is interested in a return to a pre-feminist world where women were incorrectly considered less capable due to their biology. But the nature of all cycles is that there is a time for productivity and output (think you at 11am, in the zone and kicking butt) and there's a time for slowing down and replenishing reserves (think you at 11pm, tucked up in bed and drifting off to sleep). A flower doesn't bloom all the time and neither do you.

Think of your period as the pit stop of your month, an opportunity for you to fuel up with self-love and nourishment — and probably catch up on some sleep! We all have responsibilities that can feel like hurdles we need to overcome before attending to our own needs, and these won't magically evaporate when you have your period! But if there's one week of your cycle that I encourage you to prioritize yourself, this is it. It will be easier for those of us with access to systems of support and it may require a vulnerable conversation (or two) when asking the people in your life for help. Conserving energy in your Winter can lead to more energy in your Spring and Summer, so treat it as an experiment. use your period to press your inner reset button, fill your own cup and power-up for the cycle to come.

Gut instinct

Did you know that some indigenous communities would honour women's menstruation as a time for dreaming and visioning for the tribe? Almost as if having a period was akin to becoming an oracle for a few days every month. As sensitivity increases, you have a direct line to an inner guidance that can supercharge your life and projects all month long. These creative thoughts, aka 'period ideas', can abound in your Winter phase! I often hear of women making big life decisions and coming up with solutions to problems that have been bugging them for weeks, poems and ideas appearing out of thin air, or inner callings or curiosities becoming clearer. When we slow down and tune in it's easy to feel like we're back on track, with a more solid sense of direction. The best way to access this superpower? You guessed it — taking time out (retreating to your 'period cave') and trusting yourself to reset, rest and retune.

#18

Winter's vulnerabilities

When it's hard (or not possible) to rest

Since the industrial revolution, we've forgotten that we are actually humans and not machines. We pride ourselves on being busy, thrive on getting things done, and have become obsessed with the feeling of ticking tasks off a to-do list. We expect so much of ourselves! Doing nothing feels silly and lazy and it's become a foreign concept to most of us. So if the idea of resting and relaxing when you have your period is new to you, you're not alone.

There will be plenty of reasons why pressing pause in your Winter phase will be tricky. You probably have a job to go to! Maybe even small children who need feeding! Friends who want your attention! A partner who does too! Towels to fold and trains to book and emails to send and spin class to attend and... hmmm, maybe, some of these things aren't urgent? By all means attend to the basic essentials of your life, but it's possible there are a few things on your plate that you could get away with not doing for a day or two. Next time you have your period, what would happen if you gave yourself permission to not be useful for a few hours? For 30 minutes? For 10?

Feeling blue when you bleed

The emotions that accompany the arrival of your period will be influenced by many factors. If you are trying to make a baby right now, the arrival of your period may bring disappointment and heartbreak. You may experience grief over historic miscarriages or abortions. The hormonal drop that takes us from Autumn to Winter can trigger past trauma and a sense of dark shadows and sadness arising within. Cultural conditioning, religious upbringings, menstrual shame and stigma, the feelings you have about your gender identity, body image and sexuality – yep, they're all thrown into the period pot too.

You're not weird or wrong for feeling blue when you bleed, but your period shouldn't be an ongoing source of sadness. If you experience BIG feelings as you bleed, then it's even more critical that you take exquisite care of yourself in your Winter phase. If you know that you are more susceptible to depression and anxiety in this week of your menstrual cycle, integrate this information into your mental health self-care plan with a friend, family member or mental health professional. Remember, you are not alone. (Please see #50 for mental health resources.)

#19

Blood and body love

One million, thirty-two thousand, three hundred and eighty-six. That's how many Instagram posts there are, as I type this, with the hashtag #bodylove. Scrolling through, I see a sea of smiling faces, sincere stories, belly rolls, stretch marks and power poses.

Collectively, we're learning that wobbly bits are, in fact, normal. We're learning to quit comparing our daily reflection to the glossy, Photoshopped bodies we see in magazines and in our social media feeds. It's actually just not that cool anymore to sit around with girlfriends and prod fleshy thighs and suck stomachs in. We're learning that beauty is born from diversity, confidence, and gratitude for the skin we're in! I applaud any initiative or movement that tells women:

**Your body is a gift and you are doing great,
just as you are.**

But I do wonder, where's the love for our menstrual blood? Well, technically it's menstrual *fluid*. Your period is more complex than the kind of blood that flows in your veins and arteries. Your period blood is also made up of vaginal secretions and cells of your uterine wall. But because it's always been called menstrual blood, let's stick with that.

According to the American Academy of Obstetricians and Gynecologists, your period may be considered a vital sign. Paula Hillard, M.D., Professor of Obstetrics and Gynecology, says that the menstrual cycle is a 'window into the general health and wellbeing of women' and can 'indicate the status of bone health, heart disease, and ovarian failure, as well as long-term fertility.'

Imagine if you saw the presence of your period as a monthly message from your body saying, 'Hey! You're doing great! You're well and alive and vital! Keep up the good work!'

What's more, the quantity, consistency and colour of your menstrual blood gives you direct feedback from your body, and can be an important indicator of your overall health. Keeping track of your blood helps you to get to know what's normal for you, so you can identify when and if there are any changes in your flow. Sadly, we've hardly been encouraged to exclaim, 'Thank you, body! I love you!' every time we get our period. But what if embracing our flow, in its most literal sense, could actually help us nurture a more intimate knowledge of ourselves? There's no need to throw a period party, but a grateful nod and some quiet appreciation for your vitality is a lovely place to start.

#20

Tips for period problems

If you struggle with menstrual symptoms like period pain, hormonal headaches and heavy bleeding (what joy!), you've probably scoffed at any mention of positive periods in this book — rightly so. These menstrual issues can make the Winter phase of your cycle a challenging time.

But your menstrual cycle can provide valuable insight into the inner workings of your body. Hormonal imbalances, fibroids (benign growths that develop in and around the uterus), thyroid issues, inflammatory diseases, ovarian cysts and endometriosis can all cause menstrual symptoms to surface. The hormonal shifts that occur at perimenopause and menopause can also contribute to changes in your menstrual cycle. If your period is making it difficult for you to work and live your life, please don't just shrug it off. Visit a doctor or healthcare practitioner who will listen to you and get to the root cause of your symptoms. Remember, do your own research, trust yourself, and keep tracking your periods (noting your menstrual blood loss and pain patterns). Here are some tips…

Period pain
Some discomfort on the first day or two of your period is okay — it's natural to feel your uterus contract. You may experience this as mild aches or cramping. But when the cramps are so severe that you can't go about your day as normal, the pain lasts for more than two or three days, you notice the pain is different to what you've felt before, and/or you receive no relief from over-the-counter pain relief medications, there is something else going on.

Hormonal headaches and menstrual migraines

Remember that dramatic drop in hormones (see #18) that occurs right before your period starts? That's what triggers the headaches that can affect women anywhere from two days before bleeding, right up to day 3 of your cycle. Try keeping your blood sugar levels stable by eating regular meals, staying hydrated, supplementing with magnesium, getting enough sleep and minimizing stress, alcohol, caffeine and sugar in your Autumn phase. Chart these self-care changes and observe if anything helps. If your hormonal headaches are more like menstrual migraines and the pain is not relieved with over-the-counter pain relief medication, your doctor may have more options for you.

Heavy periods

If the volume of your entire period is regularly over 80ml (13½ tsp) or you need to change your pad or tampon every couple of hours because you're repeatedly soaking through them (bleeding on furniture anyone? Yep, I've been there), then you have heavy menstrual bleeding and it's important to see your doctor for a thorough health assessment. Ask your doctor about your iron levels as heavy periods are the most common cause of anaemia for people with periods. You may find that adopting an anti-inflammatory diet (with plenty of whole grains, leafy greens, fruits, healthy fats and proteins) and minimizing your intake of dairy products, sugar, caffeine and alcohol can help.

#21

Use a menstrual cup

A menstrual cup is exactly what it sounds like – a small silicone cup that sits in the vagina, held in place by suction, that collects your blood. When it's full or you want to empty it, you simply squeeze the base and remove it. You can pour your blood down the toilet, give it a quick rinse or a wipe, and back in it goes. It's an environmentally friendly option because one cup can last for years and years. What's also handy is that the measurements on the side of most cups will tell you the volume of your flow, and because you see it in its liquid state, it makes it easier to keep track of the colour and consistency in your cycle charting practice. It can be helpful to have two menstrual cups so you can alternate them between daily cleanings (see opposite).

After using a menstrual cup for the first time, my client Miranda told me:

'It was a real moment of surprise the first time I used my cup. It was fascinating to observe the colour and consistency of my blood. I feel like my increasing levels of self-love are a result of me connecting with my blood and realizing that it's not actually gross at all, that my body isn't something to be ashamed of.'

#22

How to use your menstrual cup

Give yourself permission for your cup technique not to work right away! It can take some practice. Find somewhere private where you can relax. You'll need to fold your cup before inserting it, so follow the manufacturer's instructions. Squat or raise a leg up on the sink or toilet, or lie on your back with your legs apart and knees up.

Holding your cup in one hand, insert it into your vagina by pushing it in at about a 45-degree angle, aiming towards the small of your back. The cup should pop open — you'll know it's open by hearing or feeling a 'pop'! You can also check by running a finger around the base of the cup and making sure it's opened. Push the cup in until it's comfortable.

When do you need to remove the cup?
For your first few cycles, it'll be a bit of an experiment, but the recommended maximum time is 12 hours. On heavier days, it may be sooner. When it's time to remove the cup, you'll find it helps to bear down on the cup using your pelvic muscles, as if you're doing a poo. Insert your fingers into your vagina, find the stem, and pinch the base to break the seal. Wriggle it about a bit, slowly removing the cup, and keep it upright to avoid spillage. Empty your blood into a toilet or sink or, my favourite, pour it on your plants — they love the nitrate! Give the cup a rinse and re-insert it.

Cup care
Clean your cup at the end of each period day by sterilizing it in a pan of boiling water for a couple of minutes or by swabbing it down with rubbing alcohol. At the end of my period, I keep my sterilized cup in a cute red drawstring bag, ready for the next time I need it.

#23

Reusable pads and period undies

Billions of disposable menstrual products (including tampons, pads and applicators) are thrown out or flushed each year. But you don't have to do this. Reusable pads and period undies are, like the menstrual cup (see #21 and #22), options that are kinder to the environment and kinder to your body. No unnecessary chemicals necessary – and your wallet will thank you in the long-term too.

Some people find that period undies and reusable pads are an easier transition than going straight to a menstrual cup, because they're more familiar. Other women enjoy the feeling of 'free bleeding' that these products provide. Personally, I feel deeply satisfied when I rinse my pads and period undies out in the shower before throwing them in the washing machine. It feels like I'm living mindfully – I'm in charge of how I approach my period and how I treat my body.

What about tampons?

Your menstrual product choice is a very personal one and some people love using tampons. If that's you, all the period power to you. But for others, tampons can cause vaginal dryness and make period pain worse. If you do use tampons, make sure you go for organic, as some brands use nasty chemicals and fragrances that you may want to keep away from your most intimate area. Remember, too, that tampons have an environmental impact and while organic cotton tampons are biodegradable, they can take a long time to breakdown. It's vital to dispose of your tampons responsibly. Always throw them in the trash. Never flush them.

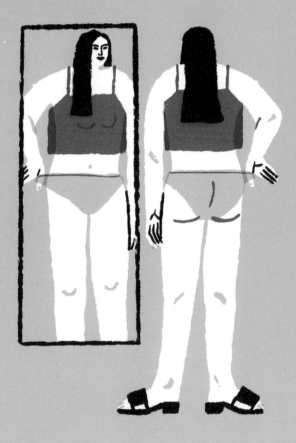

#24

Winter cycle charting Qs

Physical
What are your energy levels like when you have your period? Do they change day to day? Do you experience cramping, bloating or period pain? How about headaches? What about the quantity, quality and colour of your blood? Is there a change in your libido or sexual energy at menstruation? And your sleep?

Mental
Do you have a clear and calm headspace when you have your period? Or is this a phase where you can be more self-critical? Are your thoughts about yourself and your body more positive or negative in this phase? How about your focus and insight?

Emotional

Do you feel a sense of release and emotional refresh? Or is there more grief and sadness in your heart? Maybe all kinds of feelings at once?

Spiritual

How are your levels of self-connection in your Winter? When you slow down, what do you notice about yourself? Do you feel more intuitive here? How about a desire to be in nature or to disconnect from the material world a little?

#25

Winter cycle-syncing tips

Red-letter days
If your cycle is regular, try highlighting your upcoming periods in your calendar in RED and schedule in some downtime for yourself. With that time blocked out in advance, be discerning with what you say yes to on or around your RED days.

A gift to your Winter self
Create a period self-care box to open on cycle day 1. Fill it with treats like essential oils, dark chocolate, bath salts, an eye mask, a red shawl and your go-to menstrual products.

Winterful
Download some guided meditations, yoga nidra recordings (think yoga meets sleeping) or chilled music to your phone, and create a Winter playlist of soothing sounds. (See #50 for a link to my Winter playlist.)

Go with the flow
If you feel more intuitive or otherworldly here, be open to it. Grab a red pen and do some free-flow, uncensored journaling.

Winter visions
Just as some people create vision boards in January to guide the new year, you can do something similar for your menstrual cycle. Jot down a word or two that captures how you want to feel this cycle, some self-care intentions (for example, this cycle I'll drink more water and stick to an evening routine), and a few professional goals.

Gamechangers

Some mild discomfort on the first day or two of your period is okay and may be alleviated by:

- Sipping warm and hydrating beverages throughout the day – no, that doesn't include coffee!
- Befriending a hot water bottle.
- Taking an over-the-counter anti-inflammatory such as ibuprofen.
- Doing some gentle yoga.
- Relaxing with deep breathing. Inhale for 4 counts, hold for 7, exhale for 8.
- Gently massaging your belly with essential oils – clary sage and lavender are lovely!
- Replacing tampons with a menstrual cup, period undies or pads.

Spring

#26

Spring phase: pre-ovulation

Now we move into the pre-ovulatory phase of your menstrual cycle, a transition season, meaning that it takes you from menstruation up to ovulation. If you carved some time out in your Winter to rest, push the reset button and get clear on your priorities; you're going to be crossing over into your Spring with energy, clarity and direction.

The gradual increase of oestrogen in this phase can lend itself to a playful attitude, natural focus and increased memory retention. Just like the hours in the morning when you are waking up and getting started on your day, this Spring phase is all about re-emerging and taking action. I often hear from women that this is their favourite season in the entire cycle!

But there is some vulnerability here too. Like a rosebud gently blossoming out into the world, you haven't yet developed a thick skin. You're fresh and new in the pre-ovulatory phase, and it's important not to rush and push yourself too early. Burnout isn't sexy and if you're arriving at ovulation feeling exhausted, you might be hustling too hard in your Spring. But the truth is that your gears are shifting up, up, up. So long, period cave — you are back in the game!

THE SPRING LOWDOWN

When is your Spring phase?
Approximately cycle days 6 to 12, but this will vary depending on when ovulation occurs.

What's your body doing?
This phase of the cycle is all about preparing for ovulation and you'll begin to cross into your Spring as your period finishes. Your pituitary gland releases follicle-stimulating hormone (FSH) and your uterine lining begins to thicken, but it's oestrogen that dominates this phase. The hormonal increase of oestrogen here is linear, increasing each day, and this steady growth means that energy and serotonin levels are generally pretty high. Testosterone increases in the last few days of Spring as your body gets set to ovulate — hello strength, libido boost and energy surge!

#27

Spring's superpowers

Playfulness

Oh how wonderful it feels when the blossoms begin to appear on the trees after a cold and dark winter, and the days stretch out a little longer. There are folks frolicking in parks, smiles glued to faces, a sweetness in the air and a lighthearted energy all around. As oestrogen waxes in your body after your period ends, you might experience a feeling just like this. It's time to have fun! And let's be honest, as very responsible and grown-up adults, we can often be so busy and so boring. Spring is the moment in the cycle to tap into your inner child, and do something a little bit silly or frivolous or flirtatious. You can be very productive in this phase of your cycle, but make sure to balance it out by doing things you enjoy too.

Motivation and momentum

In your Spring, taking action can feel easy and fun! Oestrogen is growing day by day. This hormone wants to get you up and out into the world in order to find a potential mate — we are gearing up towards ovulation after all. Whether baby making is on the cards for you or not, you can ride this powerful force to bring your own goals and ideas to life. After the chilled vibes of your Winter phase, it's likely you'll feel motivated by a physical surge in energy, increased mental focus, and the feeling of potential that this new cycle holds. It's time to see exactly what you are capable of! The rising momentum of the cycle is underway, and if you can catch this wave in your Spring phase, you'll be amazed by how focused and productive you can be. Use this moment in your cycle to get the wheels in motion on creative projects and to get ahead on your to-do list.

Planning

Optimize the growing energy in the first few days of your Spring phase by sprawling out with your calendar and scheduling in the tasks, actions and steps that are going to move you closer towards your goals and dreams. From this vantage point in the cycle, you have a clear vision of what the next few weeks are going to look and feel like. What will you add to your to-do list this Spring phase? Who do you want to connect with in your Summer? When will you need more space in your Autumn? What reminders can you set now for the tender days in the cycle?

HOT TIP

Cycle-sync your schedule. Write your upcoming cycle days in red next to the date. You'll see right away whereabouts in the cycle you'll be so you can plan accordingly.

#28

Spring's vulnerabilities

When you do too much too soon

Without careful thinking or planning, it's easy to wind up in the middle of Spring feeling overwhelmed, anxious and tired. It's so easy to see why this happens! Physical energy has lifted and the fog has cleared. Plus, there's just so much to do! Try pacing yourself as you cross into your Spring phase by being discerning with what you commit to. Check in with yourself: how much physical energy and mental space do you actually have available? Remember, you're not at the peak of the cycle yet. Be careful not to deplete your resources too early, or you might wind up deficient and exhausted later in the cycle.

Self-criticism

For my client Cara, her Spring phase is when she finds herself being overly self-critical...

'This idea is awful, what on earth were you thinking?'
'You're never going to be as good as they are!'
'Try harder! Do more! You're procrastinating! You're so lazy!'

For some women, the Spring phase (particularly the first few days) can be a vulnerable time. Yet we don't demand a flower bloom faster than it does! We don't expect a two-year-old child to be able to recite the alphabet backwards. Go gently as you re-enter the world and be kind to yourself in your Spring.

#29

How to make a less and more list

A less and more list is a quick exercise to help manage 'overwhelm' and get clear on what's most important to you. You can write this list in any phase of your menstrual cycle, but I find it's helpful in Spring. There's so much possibility in this phase, but discernment is key.

1. Draw a line down the middle of a page.
2. Label the left column 'Less' and the right column 'More'.
3. What could you be spending less of your time and energy on? What's draining you or contributing to any feelings of overwhelm? Fill your 'Less' column with ideas!
4. What could you be spending more of your time and energy on? What are you craving more of? Where would you like to focus your attention during this cycle? Fill that 'More' column.
5. Circle three things from your 'Less' column that you'd like to spend LESS time and energy on this cycle.
6. Circle three things from your 'More' column that you'd like to spend MORE time and energy on.
7. What needs to happen to make these intentions a reality?

**Make the most of your Spring energy
and take action today!**

#30

Spring cycle charting Qs

Physical
When I first started charting my cycle I couldn't believe how light and energetic my body felt after my period ended. Do you feel that too? Does your appetite change? How about an increase in your libido as you approach ovulation? Any changes to the wet stuff in your knickers?

Mental
How does that rising oestrogen affect your headspace? Are you focused, curious and assertive? Or more overwhelmed and anxious? Is your inner critic trying to creep in here? Or is motivation your Spring superpower?

Emotional
Spring is a transition season, which means you're working up to something, but you're not quite there. How does that build-up feel for you? Do you feel hopeful and positive? Or restless and impatient? Is this a sensitive season for you? Or playful and fun?

Spiritual
A question I'm often asked is, 'I love the self-care stuff I do when I have my period, but how do I stay connected to myself after it ends?' Do you check in with yourself in your Spring? As the current of the cycle sweeps you up and away, what does cherishing yourself look like here?

#31

Spring cycle-syncing tips

Experiment with an idea
This is the time to take action in the direction of your goals, play with possibilities, kickstart a new habit, flirt with your dreams (or that gorgeous barista), and most importantly, give yourself permission to not be perfect as you do so.

It's time to move it, move it
This is the season to get your heart rate going. Studies show the female body can build more muscle mass in this phase of the cycle – and you should have a decent amount of energy to burn too. Lift some weights, do some laps, smash that yoga class, kickbox, run, walk, or dance your way through your Spring phase!

Let the good times roll
Plan to see close friends and loved ones and enjoy the emergence from your period cave. Take a cleansing goddess bath with fresh or dried flower petals scattered on top. Treat yourself by scheduling self-care and beauty appointments for this phase. It feels good to feel good, oui? Plus, waxing (if that's what you're into) hurts far less before ovulation.

Switch on your spontaneity switch
Spring is a time to be as foolish as you possibly can be. As an overly responsible Virgo and eldest child, the spirit of cheeky playfulness is one I actively try to cultivate in my Spring. Dancing in the kitchen to Rick James' 'Super Freak', or attempting handstands and cartwheels at your local park definitely count.

Pace yourself

Try not to rush into this phase. It's okay to pace yourself, contain your energy and maintain a bit of a buffer between you and the rest of the world. You don't have to be the star of the show right away. Save that for Summer!

Summer

#32

Summer phase: ovulation

Oh yes, we've reached the menstrual cycle summit! Characterized by ovulation, your Summer phase is the peak of your hormonal cycle. You're full of life, full of love, powerful, and expansive, and if there's one phase in the cycle where you can 'do it all', then quite possibly, this is it. Hello superwoman!

Remember that the menstrual cycle is designed to create life. Well, this is the fertile phase where it's all going down. But even if making babies isn't on the cards for you, your creativity is still flowing and your inner pleasure switch has flicked to ON. Think of your Summer mantra as, 'make art, make love, baby!'

It's easy to love Summer and many women feel at home here. But the Summer phase has a fiery energy, so no burning out, okay? If you can optimize your ovulatory energy and still save some of this sweet summery love for yourself, I promise you'll be Team Ovulation in no time.

THE SUMMER LOWDOWN

When is your Summer phase?
Approximately cycle days 13 to 19, but this will vary.

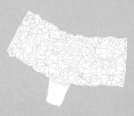

What's your body doing?

Your ovaries have got oestrogen levels surging to prepare for the release of an egg into one of your fallopian tubes. Oestrogen stimulates serotonin, which only helps to encourage the feel-good factor of this phase. Oestrogen production will peak at ovulation. The released egg then begins the journey to your uterus where your uterine lining is continuing to thicken. Follicle stimulating hormone (FSH) has risen, as well as luteinizing hormone, and we get a nice bump of testosterone here too. After ovulation, one of your ovaries will release progesterone to support a potential pregnancy. This can create a mid-Summer shift in energy as progesterone's calming and soothing qualities join the hormonal happenings.

Why is ovulation so important?

Ovulation helps us to feel great because it switches on many beneficial hormones. A successful ovulation is non-negotiable for fertility, but it is also a key player in your overall health. Ovulation is the only way your body naturally makes progesterone and the strongest oestrogen, oestradiol. These hormones help to regulate mood, energy, sleep, libido, thyroid function, skin health and more. They also contribute to growing healthier bones, cardiovascular health, and preventing some cancers.

#33

Summer's superpowers

Create, create, create!

You are a glorious, unique, creative being with so much to share with the world — what would expressing just 5% more of yourself look like? If you took action on your intentions in the Spring phase and caught the wave of momentum that oestrogen offers up, then it's easy to feel confident and on top of things in your Summer. Launching big projects and turning up the dial on your visibility may feel natural and easier here than in any other phase. It can feel like creativity is surging through every cell in your body! If it's been a while since you tapped into your creativity, use this phase to reacquaint yourself. Anything can be creative when done mindfully. Even just slowly eating a strawberry can feel ecstatic in this phase — your Summer is primed for pleasure, after all. Enjoy it.

Connect and care for others

My client Lauren, a self-employed mother of twin toddlers, knows that her Summer is when her parenting game is strong! This is the season of the cycle that she schedules special time to just be with her kids and fully embrace mothering — not to mention pre-cook and freeze meals for the family, give her partner loads of attention and get ahead on work. She doesn't expect this of herself in any other phase of her cycle (thank goodness), but in her Summer she knows it's possible. You too may feel more generous, social, and turned on to others as you ovulate. Similarly, for my client Esther, Summer is the one phase of her cycle when she's more than happy to do face-to-camera on Instagram Stories. This is because the combination of oestrogen and testosterone preps you to show up as your most confident, considerate, sexy, patient, articulate self. Work it!

HOT TIP

As you feel ovulation approaching, get clear on where you're going to focus this creative, sensual energy! What will you say 'YES' to?

A magnet for good

Think of how lovely it feels when you're absorbed in a task and 'in the flow', or when the perfect opportunity easily and serendipitously comes your way. The Summer of your cycle can feel just like this! You're at the peak of your inner rhythm here, and now you get to cruise and enjoy this natural hormonal high for a few days. The cycle is a continuum, and soon the gears will begin to shift down again, but for now, magnetism is your superpower. Remember to let your hair down, give yourself a cheeky wink in the mirror (or a sassy slap on the butt) and repeat after me, 'Everything is working out for me.'

#34

Summer's vulnerabilities

When being seen is scary

Flexing your social muscles, sharing your creative projects, turning up the visibility... this all sounds great, but what if it doesn't actually feel great? Donning a crown and strutting your stuff for all to see can leave you feeling vulnerable and exposed. For even the most confident, carefree and self-assured among us, being in the spotlight doesn't always feel incredible. If you're uncertain about how and where to direct the ovulatory creative surge in energy, it's possible this phase will leave you ungrounded and scattered. If this resonates with you, look back to your Winter phase. Are you using that time to connect with yourself and set your intentions for the cycle ahead? (See #24 and #25)

Burning out

If you arrive at ovulation feeling exhausted and drained, you're not alone. It's easy to get carried away with the highly effective energy of the Spring and Summer phases. But even a superhero has her limits. And so do you. Our society highly values productivity and output, so it's easy to see why these phases of the cycle are cherished by so many. Be careful not to give more than you have to give. Yes, these phases are about doing, creating, connecting and giving, but if you start noticing a pattern of tiredness in your Summer (and your Spring), it's a sign that you need to give some of that TLC back to yourself.

#35

Write an ovulation gratitude list

In positive psychology research, gratitude is consistently associated with greater happiness and an improvement in mental health. It's not just an Instagram trend (#blessed), but a proven and simple way to increase optimism and quality of life. As hormones peak at ovulation, you may notice a natural inclination towards sunny positivity. Ask yourself...

What do I feel thankful for today?

How and where are things working out for me?

Who in my life makes me smile?

What am I proud of achieving in this cycle so far?

Tiny things, momentous moments, people, places, ideas, mindset shifts, experiences, surprises – write it all down.

I am grateful for...

1. _____

_____

2. _____

_____

3. _____

_____

4. _____

_____

5. _____

_____

6. _____
_____
7. _____
_____
8. _____
_____
9. _____
_____
10. _____
_____

How do you feel after making your list? You might notice a boost in mood right away, but don't worry if you don't feel drastically different. The benefits of gratitude can take time to kick in. Complete this exercise in your Summer phase for the next few cycles and see what happens...

HOT TIP

This is a great way to buffer yourself for the Autumn phase ahead, when that natural positivity can feel more elusive.

#36

Summer cycle charting Qs

Physical

Run your quickest 5km? Nailing a strong vinyasa flow? Does the hormonal Summer surge give you strength and energy, or do you feel more drained by the release at ovulation? Notice any changes in your skin, hair, breasts or face? Do you feel ovulation as a twinge in your abdomen? (This sensation is called mittelschmerz!) How's your cervical fluid feeling – nice and wet? And how's that libido going?

Mental

Where do your thoughts take you in your Summer? Are you more focused on the external world, relationships and work? Do you feel on top of things here? Or does the hormonal surge at ovulation spark anxious thoughts or 'overwhelm'?

Emotional

Summer energy is creative, sensual, generous and often sexual – how does this make you feel? Does the combination of oestrogen, testosterone and serotonin make this a naturally positive time? Are you happy to be in the spotlight or does that superwoman energy feel like too much to handle?

Spiritual

Where are you expressing yourself? How do you experience pleasure and joy in your Summer phase? Do your relationships fulfil you and contribute to a sense of being loved and loving others? Self-care practices like meditation could be less appealing here, but maybe your gratitude levels feel super high.

#37

Summer cycle-syncing tips

Take a risk!
You're at the peak of your cycle, high on hormones and feeling resilient. Go all in on tasks like launching a project, speaking at an event or sharing the first draft of your latest short story with a friend.

Express yourself
How do you like to express yourself? Through clothes, taking photos or writing poetry? Making a flower arrangement for a pal? Now is the time to enjoy your creativity and share it with the world. Schedule catch up calls with friends and family, host a dinner party, see your in-laws, plan a few dates and shimmy your way through that networking event. Enjoy your people prowess!

I'll have what she's having
Plan some sensual time with yourself or a partner. For one of my clients, Sally, the ovulatory phase is fondly referred to as her 'Rabbit Week'. She and her husband try to schedule a weekend break away without their children during this phase of her cycle. Oh, yes.

Check your diary
Your Summer self will want to say 'yes' to everything and the potential to overcommit is real, so avoid doing any big planning and scheduling in this phase. This will help to reduce the need to reschedule once you get to your Autumn phase and need more chill time. Leave space in your calendar to relish in your favourite simple pleasures and keep some of this ovulatory energy for yourself.

Get cooking

Pre-cook tasty healthy meals and freeze them, ready for devouring in your Autumn and Winter phases. Because who likes cooking on their period?

Celebrate yourself

Write a self-celebration list! What have you created this cycle? Who have you connected with? How did you take care of yourself in your Winter, Spring and Summer? Write down all of your successes (tiny and big). This list will help a lot if self-doubt creeps in when you get to your Autumn phase.

Autumn

#38

Autumn: premenstruum

Like your pre-ovulatory Spring, the Autumn phase is a transition season, but this time the energy is waning. The premenstruum is the final phase in the menstrual cycle and gears are shifting down, down, down. Do you ever revert to that shiny, perfection-seeking, people-pleasing 'Good Girl'? We've all been there. Thankfully, in the Autumn phase, that melts away. You're taking off that superwoman cape — and probably your rose-coloured glasses too. That extroversion, patience and natural positivity you felt last week? There's a big chance that they've evaporated and been replaced with a primal need to be alone, in the bath, with a book and a generous glass of Malbec. On the other hand, getting weird and going raving in face paint, feathers, fluoro bike shorts and (very comfortable) trainers could be just as appealing.

The truth is that this phase of the menstrual cycle often gets a bad rap. For those who experience premenstrual symptoms like mood swings, cravings, bloating and breakouts, this phase can be a tricky and tender time. I hear you. But let me tell you this: it's also a powerful time. Many women thrive in this phase, and so can you.

THE AUTUMN LOWDOWN

When is your Autumn phase?
Approximately cycle days 20 to 27 or 28.

What's your body doing?
You'll cross into your Autumn phase around five days after ovulation occurs, and progesterone is now dominating proceedings. After the linear hormonal increase of oestrogen in your Spring phase and the peak of hormones in your Summer, in your Autumn you will experience a hormonal 'dance off' between oestrogen and progesterone. Oestrogen dips slightly post-ovulation before rising again, as progesterone continues to rise. Once your body clocks that you're not pregnant, preparation for your upcoming period begins and as you get closer to your period starting, levels of both hormones will decline. Once progesterone production comes to a complete standstill, you'll move into your Winter as you begin to bleed.

#39

Autumn's superpowers

Taking 'me time'

While the Summer woman is thrilled to wear all of the hats and cater to everybody else, in the Autumn phase your needs and desires are hankering for attention. This is the phase to set those personal and professional boundaries, claim your own turf and say, 'no, thank you'. My client Lyla, a photographer, knows that her Autumn can actually be her most productive phase, but only when she hides out in her creative cave and focuses on the work that's most important to her.

Remember, progesterone dominates this chunk of the cycle and this hormone is interested in keeping you safe and well. It can make you feel more sensitive and like you need more self-care than usual. The truth is: you do! Need some time alone? Claim it. Need more support at home? Claim that too! Just as an evening routine helps you to prepare for a great night's sleep, slowing down by taking some 'me time' in your Autumn phase is going to set you up for a much more pleasant period.

Life editing

Autumn is the season to step back, take a breath and assess: how are things really going in your life? As hormones begin their decline after the heavenly high of ovulation, you may feel less tolerant or more sensitive here – or just generally annoyed. This is a good thing! Knowing what you don't want is just as powerful as knowing what you do want. Listen to the Autumn niggles. As you edit out the stuff that doesn't feel good and true for you, it gives you more space for all that you genuinely love.

Truth telling

Lightning bolt insight, wild epiphanies and brave truth telling come alive in the premenstruum! After struggling with premenstrual frustration and anger in her Autumn phase for years, my client Rachel and I worked on understanding where she needed to set personal boundaries with her family to gain the space and independence she had been craving. For most of the cycle she was patient and happy to sweep her concerns under the rug, but in her Autumn it was a different story and she needed to draw some lines in the sand. Rachel describes it as 'knowing' that something had to change. Once these boundaries were in place, Rachel couldn't believe how quickly she began to enjoy her Autumn phase! 'I actually feel more like my true self here, like the realest and most unapologetic version of me,' she excitedly told me.

HOT TIP

Setting boundaries can often involve saying no to people. It isn't easy but, just like a muscle, it's a skill that gets stronger the more you work on it. You can do it!

#40

Autumn's vulnerabilities

Hello inner critic

We all have an inner critic that likes to point out all of the stuff we apparently suck at. Unrealistic Instagram comparisons? Feeling like an imposter? Doubting your skill set? They're just some of the inner critic's favourite (and quite predictable) tricks, and it's frequently the Autumn phase of the cycle when that voice takes the megaphone. This is the moment in your cycle to reflect, but your inner critic has a heavy hand when it comes to dishing out life advice. She (or he, or they) goes way overboard!

One of my mentors, Alexandra, a psychotherapist and leading voice in menstruality, likes to say that there's probably 5% of truth in what your inner critic says. How do you identify that 5%? Try writing some of these fears, doubts and so-called faults down. Ask yourself: does this voice remind me of someone I know? Is this feedback helpful? What is the tiny grain of truth in this criticism and what's being exaggerated? When it gets over-the-top nasty and personal ('you're a total fraud' or 'you're a terrible person'), it's time to turn the volume down.

Is anyone else exhausted?

Nina came to me with very low self-esteem and energy levels. After years of people-pleasing, she had nothing left in the tank. After repeatedly charting exhaustion in her Autumn, she knew she needed to act on it. Nina got clear on what she needed at work (more support from her team) and in her personal life (basically just more sleep), and used her Autumn truth-telling superpower to draw her lines in the sand and ask for it. Now this is the week of her cycle that she sets the boundaries that support her for the rest of the month. Slowly, her energy levels are moving back towards normal again.

Does this sound familiar to you? Look back at how you used your energy this cycle. In which phase(s) do you think you're giving more than you actually have to give? Relationships of all kinds can become strained in our Autumn, as that easygoing and patient Summer woman is gradually (sometimes swiftly) replaced with somebody with a serious need for space. So be firm, but go gently on those around you as you communicate what you need.

#41

Write an Autumn list

If there's going to be one week of the month when you feel like throwing the towel (and the entire contents of your bathroom) in, it's most likely going to be your Autumn. You might feel like quitting your job, leaving your partner or jumping on the next flight to Bali — but should you? If you're not convinced that the tropics or a divorce lawyer are actually calling you, try out an Autumn list.

Getting started
Give yourself a quiet 10 minutes or so to pause, breathe and check in with how you're feeling.

Write down anything that's frustrating you, annoying you, causing you sadness or grief, clawing at you or caging you in. It could be a relationship, circumstance or a mindset — give it a home on your list. There's zero pressure to take action on anything you write. You're just getting it out, onto the page. That's it. There's nothing else you need to do for now. Bleed on it — like 'sleep on it', but for periods!

Come back to the list in your Spring. This new cycle is the perfect time for a fresh start. Your energy and agency are rising, it's time to take those first baby steps — or seismic action! — towards positive change. But still, there's no pressure to act on anything. Clarity may take time. This is okay.

Repeat the listing process every cycle. Give more attention to recurring feelings or frustrations, and remember: trust the wisdom of your body.

#42

Self-care for PMS

Pre-menstrual symptoms can range from mild annoyance to your world being turned on its head for a week (or so). Bloating, breakouts, fatigue, tender breasts, cravings, mood swings and irritability are some of the most common symptoms. If your PMS is severe, I suggest working with a health practitioner who can help you discover the root cause of your symptoms. Thankfully, there are also simple strategies you can implement to try and ease this tension yourself.

Stress less

High levels of stress across the entire menstrual cycle, particularly the first two weeks (Winter and Spring) can increase pre-menstrual symptoms. Try incorporating a 10-minute breathing meditation into your daily routine, slashing your to-do list or deleting social media apps for a few days.

Get more sleep

An evening routine will help with this. Have an early dinner, switch your phone onto airplane mode and dim the lights in your house. Passing on booze and keeping your bedroom cool (and tech-free) can increase your quality of sleep too.

Sweat

Exercise stimulates pick-me-up endorphins, provides an emotional release and supports your liver as it processes the hormonal changes occurring in this phase. Exhausted by the thought of a HIIT workout? Go for a long walk, sauna session or a yoga class.

Sexual healing

Remember the sweet pleasures of your body via the magic of an orgasm or two... or three.

Love your liver

Like a daily car wash for your body, your liver is working hard to eliminate excess hormones (particularly

oestrogen) from your system in your Autumn. Cheer it on by staying hydrated, taking supplements with herbs (like milk thistle, turmeric and dandelion root), and cutting down on alcohol and caffeine in this phase.

Scrap sugar

Excess sugar wreaks havoc on gut health, a key contributor to hormonal harmony and mood regulation. It also disrupts natural levels of insulin and increases oestrogen. This can all lead to pre-menstrual chaos! Sweet cravings? Go for medjool dates, banana smoothies and dark chocolate.

Feel your feelings

Sometimes what we label as 'PMS' is actually our inner emotional intelligence trying to tell us that something needs to be felt. While it can be tempting to numb out with wine or food, all emotions are valid and valuable. Don't be afraid to explore the depths of yourself. Breathe into them, dance them out, talk them through, write them down and cry. Crying is your body's release valve for stress and anxiety, so embrace those Autumn tears.

When it all gets a bit too dark

The Autumn phase of the cycle can be a challenging time. I look forward to the day when our mental and emotional healthcare systems incorporate menstrual cycle awareness into their programmes and their policies. Premenstrual dysphoric disorder (PMDD) is a more severe form of premenstrual syndrome and affects 3 to 8% of people with periods. Symptoms are physical and psychological and can threaten mental wellbeing. Charting your cycle can help to identify if and when depressive and anxious feelings occur for you, and I encourage you to share these observations with a trusted loved one or mental health practitioner.

#43

Autumn cycle charting Qs

Physical
Sore boobs, bloating and breakouts? This can be an uncomfortable time – but is it for you? Do you feel a sense of 'groundedness' as progesterone increases? An increase in your appetite? How about your pain threshold and sensitivity? Do you need more sleep and rest? Do you feel like moving your body in a different way?

Mental
What's your headspace like in your Autumn? Do you bump up against your inner critic, encounter self-doubt, make demoralizing workplace or Insta comparisons, or face depression or fear? Or do you experience deep, soulful or detail-oriented thinking that propels your creativity? Is it everything all at once?

Emotional
Your Autumn will probably contain more emotional depth and variance than any other season. How do your feelings ebb and flow in this phase? Are you comfortable with your sensitive, more contemplative side? Can you face and feel the full spectrum of your feelings or is it more difficult for you to navigate your heart?

Spiritual
As you move away from ovulation, do you feel more connected to yourself and your intuition? How do you get your 'me time'? Is it by being alone, journaling or being in nature? Or a more creative expression like dancing, painting or writing?

#44

Autumn cycle-syncing tips

Details, details, details
Schedule creative tasks like editing, wrapping up a project and anything detail-oriented for this week. **Hint: you can be super productive here when distractions are kept at bay! #boundaries**

The art of saying no
Do you often cancel plans in your Autumn? Try writing 'SAVE SOME HEADSPACE!' or 'IT'S OKAY TO SAY NO' over your Autumn days. So when an invite pops up to a late night networking event or a pal asks you to help them move house on what's likely to be the last day of your cycle, you can give a considered answer.

Me time
Keep an evening (or two) free to give yourself space to do things that leave you feeling nourished and well rested — or to let off some steam. Watch a marathon of sad films or stomp about the house furiously, moving the energy through your body. A soundtrack of Erykah Badu or Adele optional, but recommended. (See #50 for a link to my Autumn playlist.)

Inner intervention
Keep an eye on that inner critic. To help prevent a spiral of negative thinking, set a reminder on your phone for your mid-Autumn phase so you'll remember to ask yourself, 'Is my inner critic chatting away?'

Clear out

Listen to those Autumn niggles and declutter your life. Maybe there's a drawer of clothes that needs sorting (I know I'm going to bleed soon when I find myself frantically cleaning out the fridge). Maybe unnecessary monthly expenses are bothering you — are you actually using that pilates membership? It could be irritation about a habit you're ready to kick or the realization that a relationship requires an adult conversation.

Protect yourself

You might choose not to launch big business projects, ask for critical feedback or, say, try your hand at stand-up comedy in your Autumn.

Winter is coming

Get ready for your Winter phase. Have you got food in the fridge? A relatively chilled few days ahead? A stash of chocolate and your heat pack ready? Called in for more support at home? Now's the time to do whatever you need to feel supported as your period approaches.

Well done! Now that you understand the seasonal phases of the menstrual cycle, let's explore some more ways to invite period positivity and power into your life.

Period
Positive

#45

Get to know your crossover days

Your crossover days are the points in your cycle that carry you from one season to the next. You may find some transitions easier or harder than others. My client Amanda finds the transition from her Winter to Spring tricky because Spring isn't her favourite season. But she loves the feeling of crossing into her Summer, as she thrives at ovulation!

Over the next couple of pages you'll find the cues that signal you're on a crossover day.

Autumn → Winter

When will I feel it? Either a day or two before you bleed, or once you begin day 1 of your cycle.

Crossover cues

- a feeling of heaviness in your womb and maybe some cramping.
- some spotting or full blood flow.
- a drop in your BBT as progesterone declines (see #38).
- a reduction in premenstrual bloating, tension or breast tenderness.
- a desire to disconnect from the world, drop your responsibilities (whether you can or not) and be alone.
- feeling a bit spacey and void-like.
- feeling bliss and relief, or feeling tense and resistant to what's coming!

Winter → Spring

When will I feel it? Around days 5 or 6 of your cycle.

Crossover cues

- you're no longer bleeding or only a tiny amount.
- feeling light and energetic in your body.
- mentally you feel clear and curious, with a sense of direction and focus.
- an increase in confidence and self-assuredness.
- a desire to connect with people and get stuck into work.
- feeling joy in your heart, or feeling some sadness if you really love having your period!

Spring → Summer
When will I feel it? Around 2 to 3 days before you ovulate.

Crossover cues
- lots of slippery 'raw egg white' cervical fluid in your knickers (see #9).
- feeling super energetic and strong in your body.
- like you want to say YES to everything!
- thinking more about your relationships.
- giving back to the people in your life.
- feeling super horny and turned on.
- a sense of natural gratitude and positivity.
- feeling positively queen-like, or feeling tender if you find ovulation to be overwhelming or tiring.

Summer → Autumn
When will I feel it? Around 5 days after you ovulate.

Crossover cues
- creamier, drier cervical fluid, and less of it.
- a gear shift down in your physical energy.
- needing more sleep.
- a drop in your libido and how 'wet' you become when aroused.
- an increase in cravings for carbohydrates and sweet foods.
- you notice you are withdrawing, feeling sensitive and needing more time alone — you can't really be bothered with anything that doesn't feel important to you.
- feeling like your 'true self', or feeling wobbly if you love your Summer and find the premenstrual week challenging.

#46

How to talk to your people about periods

Remember when the best way to get out of high school sport was telling your teacher you had your period? The world doesn't LOVE talking about menstrual cycles. But as you learn more about the way your body works and how your period affects you, it's natural to want to talk to your friends, family and maybe some of your colleagues about what you're learning. Our loved ones aren't mind readers though. You need to know your own flow and lead the way. No more replying 'I'm fine', when what you'd really like to say is, 'Could I possibly have a hug?' Or some space, or some help, or whatever it is that you need!

Explaining the menstrual cycle to someone who has never had a period
The four seasons of the cycle make sense to most because we all understand the rhythm of the year. Try this:
'You know how you feel like chilling at home when it's cold and wet outside but are positively thrilled to be out socializing when there are blue skies and the sun is shining? That's how the difference between having my period and ovulating can feel to me.'

You can also use the cycle of the day:
'Think about how you're bursting with energy mid-morning and love getting stuck into work, but then feel like napping in the afternoon. Then you sleep and rest and wake up again with more energy. That's similar to what I experience over the course of my cycle.'

Reaching out for cycle support

I asked my fab Instagram community what kinds of cycle support they asked their partners and family members for. Here's what they said:

- to give me time alone, help with house decisions and cook dinner.
- being able to talk openly about my period without any shame or groans.
- to recognize that some weeks are just a bit harder than others.
- to take the kids so I can listen to my own thoughts and rest.
- refilling my hot water bottle and having one ready when I get home.
- back rubs and lots of cuddles.
- my kids aren't allowed to call 'Muuuummmmm' across the house. Dad is on duty!
- patience and kindness, even when they don't fully understand what's going on.
- to run me a bath with essential oils and tell me not to move.
- chocolate!!!

#47

What does the menstrual cycle have to do with the moon?

The words 'menstruation' and 'menses' are derived from the Latin 'mensis' (month), which relates to the Greek word 'mene' (moon). In many ancient cultures, the moon was associated with goddesses such as deities Selene (Greek), Chang'e (Chinese), Luna (Roman), and the Triple Goddess archetype in Wicca tradition. Native American Yurok women referred to menstruation as 'moon time' and would take days off from tribal responsibilities at the new moon to bleed – they were considered to be at the height of their powers during their periods.

If you look at the moon tonight you'll see that it's in one of four primary phases, just like anyone with a menstrual cycle.

The lunar cycle looks like:
• New moon (the sky is dark and the moon can't be seen)
• Waxing moon (the moon appears to grow)
• Full moon (bright and full illumination)
• Waning moon (the moon appears to decrease in size)

Sound familiar? Some consider the new moon to reflect the Winter phase, the waxing moon the Spring, the full moon the Summer, and the waning moon the Autumn. And yes, the lunar cycle takes about 28 or 29 days to complete one full cycle. It can be fun to chart your cycle alongside the lunar phases, however they correspond to your seasons, but whether or not you feel a connection to the moon is completely personal.

#48

As to Qs I get asked a lot

Q: I'm on the pill! Do I still experience the four seasons?
A: If you're on hormonal contraception (the pill, hormonal IUD, implant, Depo Provera shot or Nuva Ring), you don't ovulate and therefore won't experience the ebb and flow of oestrogen and progesterone over a cycle. Some women on hormonal contraception tell me they do feel the seasons of the cycle, while others tell me they feel as if they're in one season all of the time. I say, trust how you feel.

Q: I'm pregnant (or breastfeeding) and miss my cycle! Do I have one?
A: Some consider the first trimester of pregnancy to be like an extended Winter phase, the second trimester as an extended Spring, the third trimester as a Summer phase and postpartum as Autumn. If you miss charting your cycle, borrow the moon's. Simply follow the lunar phases, charting as if they are your own (see #47). I also recommend this method if you're in or beyond menopause.

Q: How do I plan for my Winter when I don't know it's coming?
A: The menstrual cycle is variable by nature. I still can't predict exactly when my period is going to arrive. Charting ovulation will help and a BBT temperature drop in your Autumn is a clear sign you're about to bleed. But sometimes surprise periods happen! Some plans can change but when they can't, flexibility and self-kindness is key.

Q: My cycle is only 24 days. How do my seasons work?
A: Winter is as per normal. Let's say you ovulate around day 12, meaning you have a slightly shorter Spring and cross into Summer around day 10. You'll move into Autumn around days 17 or 18.

Q: Is a 26-day cycle and then a 29-day cycle something I should be worried about?

A: Not at all! Both cycles are of a normal length and that's a normal amount of variance.

Q: My cycle is over 45 days long. How do my seasons work? Is this something to be worried about?

A: Long cycles (over 40 days) usually signal a lack of regular ovulation, and your body may be attempting to ovulate multiple times before successfully releasing an egg. If your cycle is 45 days long, you'll be ovulating from day 29 onwards, which means you'll have a longer Spring phase. This might not always be the case though. Simple lifestyle tweaks like minimizing stress, getting enough sleep and eating whole foods can often help to bring cycles back into balance, but remember, flagging a longer cycle with your doctor is an important step in figuring out what is preventing ovulation from occurring.

Q: I'm in perimenopause and my cycles are changing — they're longer/shorter/heavier/lighter. How can I chart this?'

A: Perimenopause, the transition leading up to menopause, usually begins in a woman's early to mid 40s. Cycles can lengthen, shorten or become irregular. Periods will change too. Spring might start to feel like Autumn and your Summer could last for just a few days. Charting ovulation will be helpful. Consider herbal or therapeutic support during this transition — this moment in the female hormonal life journey is one we don't talk about enough!

#49

A menstrual manifesto

In our busy modern world, having a menstrual cycle and trying to live by it can feel as if you're trying to squeeze a very large square peg into a very small round hole. How does one actually live by their cycle and embrace their period and menstruality in the real world?

The best secret to have up your sleeve (and this is one of those secrets I want you to share with everyone you know) can be summed up in three words: awareness, acceptance and action.

- You're aware of your inner rhythm and how your period and cycle affect how you feel.
- You accept this ebb and flow, softening into the truth of it.
- You take action with this information, syncing your life where you can around your inner seasons and committing to working your period power for good.

Nurturing an intimacy with your cycle is going to deliver gold. Over time, menstrual cycle awareness cultivates self-authority, where YOU become the leading decision maker in your life. You know where your strengths in the cycle lie and how to work these superpowers, optimizing them at just the right time. You also know where your more tender days are and you can take care of these vulnerabilities, knowing just what you need and when. Finding your flow is about embracing your cyclic nature and recognizing that there is a time for productivity and output, but there is also a time for rest, pleasure, sensuality, play, beauty, community, being with loved ones and having time to just BE.

I suggest getting some period pals on board too. Can you and your friends commit to helping each other out on the first few days of your cycle when you can? What about checking in with colleagues as to which season they're in and syncing projects? Choosing not to hide your tampon on the way to the office bathroom or simply speaking openly about your cycle might give you a huge boost in period confidence!

I telepathically fist-pumped actress Jennifer Lawrence when she told *Harper's Bazaar* that the motivating factor behind her 2016 Golden Globes dress was that 'awards season is synced with my menstrual cycle and this dress was loose at the front. I'm not going to suck in my uterus. I don't have to do that.' No J-Law, you do not have to suck in your uterus (literally or figuratively), and nor do any of us.

Mostly, I hope you recognize that having a period doesn't have to hold you back. All of this is really just about having the awareness to be your best self. Living life by our cycle doesn't mean that we can't lift weights in our Autumn or take some alone time in Summer. We can do anything we bloody well like. Menstrual cycle awareness just gives us an added edge, so we know when to push, when to pull back, and when to be a bit kinder to ourselves. Because that, friends, is knowing your flow.

#50

Period positive resources

Where to from here?
Download your free cycle charting resources at:
clairebaker.com/50-things
Learn more about my online courses at **clairebaker.com/courses**
Find seasonal playlists at **clairebaker.com/period-playlists**

Menstrual products
Menstrual cups: Mooncup, Juju, OrganiCup, Saalt.
Period underwear: Thinx, Modibodi, WUKA.
Reusable pads: EcoFemme, Lunapads, Hannahpads.
Disposable tampons & pads: Tsuno, Grace & Green, TOTM.
Fertility charting: Daysy Fertility Tracker, Pen & Paper Fertility.
Cycle tracking apps: Clue, Kindara (use with a BBT thermometer).
Moon tracking app: The Moon

Further reading
Beautiful You by Nat Kringoudis
Code Red by Lisa Lister
Period Power by Maisie Hill
Period Queen by Lucy Peach
Period Repair Manual by Lara Briden
Red Moon by Miranda Gray
The Billings Method by Evelyn Billings
The Pill: Are You Sure it's for You? by Jane Bennett & Alexandra Pope
Wild Power by Alexandra Pope & Sjanie Hugo Wurlitzer
WomanCode by Alisa Vitti

Further support

For diagnosis and treatment of conditions such as endometriosis, polycystic ovary syndrome (PCOS), fibroids and premenstrual dysphoric disorder (PMDD), start by making an appointment with your GP. You may find complementary support through a holistic practitioner such as a naturopath, herbalist or qualified acupuncturist who uses menstrual cycle awareness in their toolkit. The Association of Naturopathic Practitioners (theanp.co.uk) and The British Acupuncture Council (acupuncture.org.uk) will aid your search. To avoid or achieve pregnancy naturally, I recommend working with a trained fertility awareness teacher.

For mental health support, start with an appointment with your doctor. They may make a diagnosis, offer medication or treatment, or refer you to a mental health specialist. Trained psychologists, coaches and therapists are an option, as well as helplines such as Samaritans.

Share your period positivity!
Find me on Instagram at
@_clairebaker_ and use the
hashtag #50thingsaboutperiods
to tell me how you're going
with your flow.

Acknowledgements

It wouldn't have been possible to write this book without the hundreds of women I've had the honour of coaching. Thank you for trusting me. I will be eternally grateful for the teachings of Alexandra Pope and Sjanie Hugo Wurlitzer of Red School, as well as Dr. Kerry Hampton. Thank you to Valeria Huerta and Brigid Moss for believing in this book from the beginning. To the wonder-team at Pavilion, especially Katie Cowan and Krissy Mallett, thank you for bringing the menstrual magic to the page. To Gen, Lacey, Laura, Mel: you read drafts, shared stories and gave golden feedback — I appreciate you! Clara Bitcon-Bailey, your generous insights elevated my message, and Rozalina Burkova, your vibrant illustrations brought this book to life. To Alex, who has been on Team Period since our very first date, I am crazy grateful for your love and support. Finally, to my parents, thank you for cultivating a period positive home to grow up in — it's made the world of difference.

First published in the United Kingdom in 2020
by
Pavilion
43 Great Ormond Street
London
WC1N 3HZ

Illustrations by Rozalina Burkova

ISBN 978-1-911641-64-3

A CIP catalogue record for this book is available from the British Library.

10 9 8 7 6 5 4 3 2 1

Reproduction by Rival Colour Ltd, UK
Printed and bound by Toppan Leefung Printing Ltd, China

www.pavilionbooks.com

The information in this book is not intended as a substitute for professional medical advice and/or treatment. If you are pregnant, breastfeeding or suffering from any medical conditions or health problems, it is recommended that you consult a medical professional before following any of the advice suggested in this book. The publisher, author, or any other persons who have worked on this publication, cannot accept responsibility for any injuries or damage incurred as a result of the information or advice contained in this book.